Computing and Systems Technology Division

Held at the 2023 AIChE Spring Meeting and 19th Global Congress on Process Safety

Houston, Texas, USA
12-16 March 2023

ISBN: 978-1-7138-7927-5

Printed from e-media with permission by:

Curran Associates, Inc.
57 Morehouse Lane
Red Hook, NY 12571

Some format issues inherent in the e-media version may also appear in this print version.

Printed with permission by Curran Associates, Inc. (2023)

AIChE
120 Wall Street, FL 23
New York, NY 10005-4020

Phone: (800) 242-4363
Fax: (203) 775-5177

www.aiche.org

Additional copies of this publication are available from:

Curran Associates, Inc.
57 Morehouse Lane
Red Hook, NY 12571 USA
Phone: 845-758-0400
Fax: 845-758-2633
Email: curran@proceedings.com
Web: www.proceedings.com

TABLE OF CONTENTS

46a Effect of Operational Conditions on the Safety (thermal characteristics) of Cylindrical Lithium-Ion Battery

Sankhadeep Sarkar, Mahmoud El-Halwagi, Faisal Khan, Md Tanjin Amin

Operational factors (e.g., current rate, external temperature) influence the electrochemical reaction rate and heat generation in a lithium-ion battery. Standalone electrochemical models (EM) limit the estimation accuracy by considering only fixed electrochemical parameters.

Thus, an EM coupled with a thermal model (P2D-T) is proposed to understand battery behavior accurately. In the coupled model, the heat generation rate from the P2D model is introduced to the thermal model where the temperature-dependent electrochemical parameters (e.g., rate constant, diffusion coefficient, ionic diffusivity) are updated using Arrhenius expression and returned to the P2D model. The P2D-T is employed for a cylindrical type $LiFePO4$/Graphite battery design, simulated at different current rates (1C to 5C).

This study aims to demonstrate the relationship (e.g., the sources) between non-linear heat generation terms inside the batteries (e.g., reaction, entropic, ohmic). These relationships help to quantify heat generation term, evaluate thermal sensitivity based on different charge rates. This study provides an understanding the intrinsic battery behavior and thermal sensitivity will help the researchers to design robust safeguards against thermal runaways.

46b Computational Modeling for Risk Analysis to Failure and Stability of the DAM

Igor Melo

The present study aims to demonstrate modern risk analysis solutions based on computer technology and equation modeling to effectively predict dam infiltration problems, their consequent collapses and effectively assist vulnerability studies. In the 2D/3D modeling studies through the PLAXIS software, data of different characteristics can be presented: the permeability and the hydraulic resistance of the soil change with the moisture content, the special properties of unsaturated soils. Risk Analysis for operations with adverse weather conditions and maintenance needs can also be added to the analysis, which provides an adequate analysis in the choice of maintenance models and methods. Theoretical studies and manual input data are compared, with data parameterized by recognized software and used in investigations of dam collapses as occurred in Brazil, in the cities of Brumadinho and Mariana. The result is a technical and effective basis for quantitative and numerical analyzes to determine the best intervention alternatives in operation and maintenance, with a view to choosing the lowest and/or most controlled risks.

46c Path-Sampling and Dynamic Risk Assessment for Rare Abnormal Events – Part II – Alarm Rationalization Strategies

Vikram Sudarshan, Warren Seider, Amish J. Patel, Jeffrey E. Arbogast, Ulku Oktem

Chemical manufacturing processes can pose serious hazards. Thus, safety considerations play an important role in their design. Indeed, to minimize the risk of catastrophic accidents, which can result in loss of human life or major environmental impacts, extensive instrumentation such as control systems, alarms, and automated Safety Instrumented Systems (SIS), are routinely employed in chemical processes. While such efforts are successful in mitigating the most likely postulated abnormal events, it is challenging to anticipate and mitigate the effects of infrequent, un-postulated, abnormal events in real-time. Importantly, such rare safety events, which have not been considered in plant design, can lead to the most severe consequences.

In this research, novel, multivariable alarms and safety systems are introduced using process modelling and path-sampling for un-postulated abnormal events resulting from perturbations in one or more process variables. Forward-flux sampling (FFS), developed to discover rare molecular dynamics pathways, is applied (Allen et al., 2009). For a proportional-only controlled exothermic CSTR approximate process model with perturbed feed concentration, the FFS algorithm is applied to identify rare trajectories between high- and low-conversion steady states, with key process variables saved at various 'crossing points' (Sudarshan et al., 2021). Then, committor probabilities, p_B, are computed at each crossing point (Borrero and Escobedo, 2007), yielding a mathematical model that expresses p_B as a function of the key process variables; i.e., the reactor temperature, cooling-water flow-rate, and cooling-water temperature (Sudarshan et al., 2022). Alarm thresholds, i.e., *L-low*, *LL-low-low* and *ESD* (emergency shutdown), and their corresponding safety actions, are suggested by computing the critical ranges of the process variables, given the p_B ranges for every alarm threshold.

Next, as described herein, using alarm rationalization strategies, the acceptability of every alarm threshold is evaluated, with the alarm thresholds and/or safety actions modified accordingly, based on key statistical metrics — this ensures that every alarm is a quality alarm, and its safety action is justified appropriately. Lastly, their real-time performance is evaluated using dynamic risk assessment, in which, the risk associated with the alarms and safety systems is analyzed by estimating the failure probabilities of the safety systems. These failure probabilities are estimated based on multiple dynamic simulations for the process, inclusive of control, alarms and safety systems, and the low-variance probability distribution constructed using these estimated failure probabilities is referred to as the informed prior distribution. Given the unavailability of safety system failure data, the likelihood distribution is constructed using assumed data, following which, the posterior distribution is constructed using Bayes' Rule. Expectedly, the posterior distribution constructed using the informed prior has a much lower variance compared to one constructed using a flat prior — the latter depends entirely on the likelihood distribution having a larger variance, and hence, is unreliable.

Keywords: Un-postulated Abnormal Events, Forward-flux Sampling, Committor Probabilities, Alarm Rationalization, Dynamic Risk Assessment

References:

Allen, R.J., Valeriani, C., Rein Ten Wolde, P., 2009. Forward flux sampling for rare event simulations. Journal of Physics Condensed Matter 21.

Borrero, E.E., Escobedo, F.A., 2007. Reaction coordinates and transition pathways of rare events via forward flux sampling. Journal of Chemical Physics 127.

Sudarshan, V., Seider, W.D., Patel, A.J., Arbogast, J.E., 2021. Understanding rare safety and reliability events using forward-flux sampling. Computers and Chemical Engineering 153.

Sudarshan, V., Seider, W.D., Patel, A.J., Oktem, U.G., Arbogast, J.E., 2022. Path Sampling and Dynamic Risk Analysis for Rare Abnormal Events – Part I – Design of Multivariable Alarms and Safety Systems for an Exothermic CSTR – in preparation.

46d Robotics and AI to Improve Process Safety Performance

Elbert van der Bijl

Robots have been used for decades in discrete manufacturing and warehousing; can they be used just as effectively for the process industry, especially with respect to improving safety?

Technology has advanced in recent years to allow robots to move autonomously and collect data. Several robotic companies have developed such robots with various mobility capabilities and sensor options. Sensors include those that most humans inherently possess in addition to superhuman ones, such as thermal imaging and gas sensors.

With the addition of AI, the data collected from the sensors can be used to evaluate anomalies and trends, thus preventing equipment failure and accidents. In addition, robots perform tasks repeatedly without distraction and human-error thus bringing safety to the forefront. Lastly, they also can be used for work that is dangerous, thus removing humans from potential threats.

63a Explaining the Wetting Stability and Robustness of the Cassie-Baxter State on Reentrant Micropatterned Surfaces Using Theoretical Analysis and Continuum Simulations

Rahagir Ridwan Anik, Kalen Baker, Ping He

Surface wettability has been a long-studied research topic and worked as a solution to a broad spectrum of engineering problems. In this work, a theoretical Gibbs energy model has been developed for a series of reentrant micropatterned surfaces to explore the phase transitions of wetting states. The energy barriers from the Cassie-Baxter state to the Wenzel state and vice versa are theoretically computed, aiming to design more robust superhydrophobic surfaces. A three-dimensional wetting process of a 3-μL water droplet was simulated in three different modes using OpenFOAM. First, at zero gravity, the droplet was initialized in the Cassie or Wenzel state on the micropatterned surface and was relaxed to a steady state. Second, using a controlled body force, the droplet was pushed from the Cassie to Wenzel state or pulled from Wenzel to Cassie state, respectively, to estimate the energy barriers of the two transitions. Finally, the droplet impacts on the micropatterned surfaces were simulated at different initial velocities. The comparisons showed good agreement between our theoretical model and continuum simulations. This study has also revealed the mechanistic explanations for the robustness of the Cassie state on the reentrant micropatterned surfaces, i.e., the contribution of the sharp increase in the Gibbs energy, which will help the design of more robust superhydrophobic surfaces.

63c A Simultaneous Material Screening and Process Optimization Approach for Carbon Capture Using Machine Learning

Suryateja Ravutla, Sun Hye Kim, Héctor Octavio Rubiera Landa, Matthew Realff, Fani Boukouvala

Recent advances in High-fidelity (HF) simulation-based optimization have led to the development of many algorithms that can handle high-dimensional spaces with mixed-integer inputs. However, there are challenges that remain when the inputs consist of both continuous and discrete variables and the sampling costs are limited. As an alternative, surrogate models, or Machine Learning (ML) models are employed to model HF simulation data and expedite the optimization search. Due to the complexity associated with dynamic HF simulations, many researchers have employed surrogate modeling to relate the inputs and the target outputs produced the simulation (Kim & Boukouvala, 2020) (Xiao et al., 2021).

One case study where this type of mixed-integer, simulation-based optimization problem arises is in design of adsorption processes for carbon capture using solid adsorbents. Recent studies show that the choice of adsorbents strongly affects the process performance, implying process optimization and adsorbent selection should be considered alongside (Subramanian Balashankar & Rajendran, 2019). This requires formulating a complex simulation-based optimization problem containing partial differential algebraic equations and a large search space with both continuous and discrete variables. In this study, we present a ML-based approach for simultaneous process optimization and adsorbent selection. Firstly, we propose a technique to design modular CO_2 capture systems for coal-fired power plants using Vacuum Pressure Swing Adsorption (VPSA) coupled with thermally modulated fiber composite adsorbents. We then use surrogate-based optimization to determine the optimal adsorbent as well as process conditions to meet at least 90% recovery and 95% purity constraints (Kim, Landa, Ravutla, Realff, & Boukouvala, 2022). We compare feasibility and computational cost savings of simultaneous optimization of adsorbent selection and operation using surrogate-based optimization versus a brute force optimization of operation for each material separately. For this study we explore a pool of 75 different adsorbent materials and 5 process conditions for optimization. Finally, we use the large amount of local-optima collected by our solver and utilize ML techniques to develop "adsorbent-to-optimal process" correlations. Specifically, we will show that using principal component analysis we can generate a reduced-space that represents the material properties, while coupling that with process inputs we can build classification and regression models to predict the feasibility and process outputs of different materials.

References

Kim, S. H., & Boukouvala, F. (2020). Surrogate-Based Optimization for Mixed-Integer Nonlinear Problems. *Computers & Chemical Engineering.*

Kim, S. H., Landa, H. O. R., Ravutla, S., Realff, M. J., & Boukouvala, F. (2022). Data-driven simultaneous process optimization and adsorbent selection for vacuum pressure swing adsorption. *Chemical Engineering Research and Design, 188*, 1013-1028. doi:https://doi.org/10.1016/j.cherd.2022.10.002

Subramanian Balashankar, V., & Rajendran, A. (2019). Process Optimization-Based Screening of Zeolites for Post-Combustion CO2 Capture by Vacuum Swing Adsorption. *ACS Sustainable Chemistry & Engineering, 7*(21), 17747-17755. doi:10.1021/acssuschemeng.9b04124

Xiao, J., Mei, A., Tao, W., Ma, S., Bénard, P., & Chahine, R. (2021). Hydrogen Purification Performance Optimization of Vacuum Pressure Swing Adsorption on Different Activated Carbons. *Energies, 14*(9), 2450. Retrieved from https://www.mdpi.com/1996-1073/14/9/2450

63d The ASSF Tool – an Automated Systematic Synthesis Framework for Determining Feasible Reaction Pathways Prior to Extensive Design Procedures

Jaimee Jugmohan, Josias Van Der Merwe, Mulenga Bwalya

With reactor unit operations considered to be the core of any chemical process, a considerable amount of design resources has been focused on optimizing its associated operating conditions, flow rates and catalyst composition, with the sole focus of this optimization centering around its economic and environmental feasibility. Therefore, when developing and analyzing existing, and new chemical process structures, the critical step involves determining the most efficient reactive system. While there may be a number of feasible pathways, there is always a possibility that a novel reaction system, or combination thereof, could potentially improve the efficiency of the process, or lower its operational costs.

However, with the proliferation of reaction pathways focused on the production of a single product, it has become increasingly difficult and time-consuming to develop economically and environmentally sound manufacturing processes. Whilst a novel, catalytic reaction pathway may exhibit feasibility on a laboratory scale, this may not hold true when scaled to an industrial level, leaving the potential plant application of these pathways with much uncertainty. One of the major concerns with scaling up a chemical reaction to an industrial scale, lies in the cost of unit operations required to meet the reaction's operating conditions, as well as the necessary separation conditions. These downstream processing (DSP) techniques are of particular importance to the feasibility of a process, as it determines the purity achieved from the process structure, and its associated market value.

Additionally, to increase the feasibility of a process, it is possible that the product of one reactive system be treated as an intermediate in the overall process, thereby allowing for the production of a secondary product with a greater economic value. Whilst this expansion of the process boundaries cannot guarantee a more economically and environmentally feasible solution, the possibility of such cannot be negated, and is often overlooked during the initial stages of process design.

The Automated Systematic Synthesis Framework tool (ASSF tool) developed and presented herein, is a novel, rapid screening tool that bridges the gap between the analysis and scale up of laboratory systems. This tool, focused on the automated analysis of a reaction pathway database, considers the economic and environmental impacts of a process design by; (1) providing generic process structures, (2) determining the cost and environmental implications associated with the scale up of a reactive system, and (3) systematically searching for, and identifying, the possibility of pairing reaction systems, as consecutive reactor vessels in a chemical plant, attempting to increase the overall feasibility of the process structure.

The uses of the novel ASSF tool, presented herein, include:

- Quick, automated simulations of chemical plants given only the reaction pathway involved in the process,
- Determining the likely optimum operating condition range of a reaction pathway, given experimental data,
- Systematically designing and analyzing potential chemical plants associated with a reaction pathway database, accounting for potential reaction pathway combinations, and
- Reducing the potential search space of the reaction pathway database using economic and environmental search space reduction criteria.

Whilst the ASSF tool is not designed to develop novel reaction pathway systems, it is a useful analytical tool that speeds up the search for the most promising, and suitable, reactive systems. It is also able to highlight bottlenecks, and thus aid in strategic decisions and allocation of resources. The advantage of the ASSF tool in the existing engineering software space lies not only in its automated approach, but also in its simplified, generic design that minimized the level of detail required to analyse a reactive system/s, only requiring the balanced stoichiometric equation and the reaction's operating conditions. This improves the efficiency of the tool, whilst increasing its usability in conceptual design stages, where streamlined and limited input data is available.

Once the reactions are inputted in a guided, predefined database format, the ASSF tool systematically searches for any potential reaction pathway combinations, registering each combination as a separate case study for analysis. Each reactive system/case study then undergoes a basic, generic process design, costing the unit operations required in, and leading up to, the reaction and separation systems. These plant structures are then analysed using economic and environmental criteria, with the final, reduced search space being ranked in order of its expected industrial feasibility.

It is important to note, however, that the ASSF tool is not designed as a replacement for existing chemical engineering software, but rather acts as a precursory, screening tool, with the ability to automatically narrow down large reaction pathway search spaces in a fraction of the time that would have previously been needed if these pathways were individually simulated in the existing software packages. Once the search space has been reduced and ranked in order of their predicted feasibilities, the user can utilize intricate software platforms, such as Aspen ® Plus, to perform a detailed design of its process structure. This allows for more efficient use of design resources, with users now only needing to focus on the intricate design of pathways that are likely to exhibit feasibility on an industrial scale. This is also beneficial in avoiding committing expensive resources from the onset.

To ensure the accuracy of the results obtained from the ASSF tool, the developed tool was verified against published simulation platforms, such as CHEMCAD ® and the Web-based Economic Cogeneration Modular Program (W-EcoMP). These comparisons show a minor deviation in capital costs of less than 6%, which is well within the ±30% range for preliminary design procedures. Furthermore, the ASSF tool was tested with a reaction pathway database developed by Jugmohan, et al. (2020), focused on the production of methanol and dimethyl via the hydrogenation of carbon dioxide. This study noted an analysis period of 10 months. This lengthy analysis period was due to the manual design and transfer of data between four intricate software platforms. Using the ASSF tool, the database was completely analysed in under 5 hours, concluding an identical final search space of 5 potentially feasible reactive systems. This application reiterates the use and uniqueness of the ASSF tool in identifying promising reactive systems at the early onset of the design procedure, allowing resources to be more focus driven.

Optimum Design of Blue Hydrogen Production System Using Steam Methane Reforming with CO_2 Capture and Storage (CCS) Process Utilizing Liquefied Natural Gas Cold Energy

Hyeon-won Jeong[1], Khanh Vinh Nguyen[2], Shu Wang[3], Ricard Gutfraind[3], Ruichang Xiong[3], Jerry Wareck[3], Samantha Neilsen[3], Yong-kwan Kim[4], Duk-young Hwang[4] and W. Jaewoo Shim[1], (1)Chemical Engineering, Dankook University, Yongin-si, Korea, Republic of (South), (2)Charles Darwin University, Darwin, NT, Australia, (3)Aspen Technology Inc., Bedford, MA, (4)Technology Research Center, Hyundai E&C Co., LTD, Seoul, Korea, Republic of (South)

Hydrogen is an important alternative energy fuel source to help reduce emissions in accordance with the Paris Agreement. It is mainly produced from natural gas by the steam methane reforming (SMR) process in which carbon dioxide is generated as a by-product. This carbon dioxide should be captured and liquefied after which it can be stored. In the first part of our study, an optimized process for 20 ton/day scale hydrogen production is proposed. This process involves capturing and liquefying carbon dioxide from the flue gas and SMR product streams using the amine-based capture process. Liquefied natural gas (LNG), which is already used as both reactant and fuel for the SMR process, is further utilized to liquefy the CO_2 in a cryogenic heat exchanger. In the second part of this study, two types of heat exchangers are evaluated for this task: Shell and Tube and Plate-Fin. For each exchanger type, a full design is performed considering rigorous modeling of thermal and hydraulic performance, conductive distribution in axial direction, flow instability, and thermal characteristics of metals. An economic evaluation for each design is performed for the equipment, installation, maintenance, and utility costs and the results are compared. Aspen HYSYS V12 is used for the process simulations and Aspen Exchanger Design and Rating (EDR) is used for rigorous heat exchanger design and modeling.

Acknowledgement: This research was supported by the program of CCU3050 through the National Research Foundation of Korea (NRF) funded by the Ministry of Science, ICT (grant number: 2022M3J2A1063796 and 2022M3J2A1063788)

78c CFD Simulation Analysis for Refinery Piping Detail Design Optimization

Minkyu Han, Krishnadash S. Kshetrimayum

With the recent improvement in computing power, the fields to which CFD analysis can be applied also increasing. Traditionally, refinery piping design has also relied on the standard suggested by the international code and the experience of engineers, but more detailed designs are now possible through CFD analysis. In this presentation, a case in which the reference pipe was optimized and designed by CFD analysis was included. The examples shall include fluid distribution study for optimization of Air Cooler Heat Exchanger piping configuration, cooling effect study at de-superheater downstream piping and fluid pattern study about pressure drop and reaction force according to the rotation angle of the valve disc.

AUTHOR INDEX